BOXER BOOKS Ltd. and the distinctive Boxer Books logo
are trademarks of Union Square & Co., LLC.
Union Square & Co., LLC, is a subsidiary of Sterling Publishing Co., Inc.

Text © 2024 Boxer Books
Illustrations © 2024 Pintachan

All rights reserved. No part of this publication may be reproduced, stored in a retrieval system, or transmitted in any form or by any means (including electronic, mechanical, photocopying, recording, or otherwise) without prior written permission from the publisher.

First published in Great Britain in 2024.

ISBN 978-1-4547-1171-1

A catalogue record of this book is available from the British Library.

For information about custom editions, special sales, and premium purchases, please contact specialsales@unionsquareandco.com.

Printed in China
10 9 8 7 6 5 4 3 2 1

03/24

unionsquareandco.com

Spring Street™ Series created by David Bennett
Written by Sasha Morton
Illustrated by Pintachan
Series editors: Sasha Morton and Leilani Sparrow
Series consultant: Mary Anne Wolpert, Cambridge University

DINOSAURS
Contents

Walking tall 6
Among the first 8
Stegosaurs 10
More stegosaurs 12
Frills, horns and beaks 14
Small but speedy 16
Walking plant-eaters 18
Large ornithopods 20
Did they eat-it-all-osaurs? 22
Prosauropods 24
Super-sauropods 26
Beware the titanosaur 28
Jaws, claws and teeth 30
Here comes trouble 32
Early birdies 34
Flying high 36
On the wing 38
Below the surface 40
Plesiosaurs and mosasaurs 42
Mega-sharks and turtles 44

Spring Street

Walking tall

1 Stegosaurus was a slow-moving plant-eater with a spiked tail to scare off predators.

2 Triceratops was a plant-eater. It used its horns to defend itself from meat-eating predators like the Tyrannosaurus. How many horns can you count?

3 Tyrannosaurus rex was the most powerful of the meat-eaters, with 60 pointy teeth and jaws that could chomp through bone.

Dinosaurs walked the Earth hundreds of millions of years ago – that's a very, very long time! These reptiles hatched from eggs and had straight back legs that let them stand upright. **These are five of the most well-known dinosaurs: is your favourite here?**

4 Diplodocus had a long neck to reach high plants and to help it drink water.

5 Coelophysis was one of the earliest dinosaurs – it was small but fast!

1 **Stegosaurus**
(say: *steg-uh-SAWR-us*)

2 **Triceratops**
(say: *try-SER-uh-tops*)

3 **Tyrannosaurus rex**
(say: *ty-ran-uh-SAWR-us REX*)

4 **Diplodocus**
(say: *dip-lo-DOCK-us*)

5 **Coelophysis**
(say: *see-LO-fi-sis*)

Among the first

1 Chindesaurus had a long whiplike tail and was around 12 feet (4m) long – that's about as long as a car. It lived in the southwestern United States.

2 Herrerasaurus lived 228 million years ago in Argentina.

The earliest dinosaurs were carnivores, which means they were meat-eaters. Did you know the *-saurus* part of any dinosaur's name means "lizard"?

3 Staurikosaurus lived around 220 million years ago in Brazil.

1 **Chindesaurus**
(say: *chin-duh-SAWR-us*)

2 **Herrerasaurus**
(say: *huh-rare-uh-SAWR-us*)

3 **Staurikosaurus**
(say: *stawr-ik-uh-SAWR-us*)

Stegosaurs

1 Ankylosaurus was slow-moving and the size of a military tank.

2 Dacentrurus was the first stegosaur ever discovered.

3 Emausaurus lived in Germany around 180 million years ago.

4 Gastonia lived in North America over 127 million years ago.

Stegosaurs (say: *STEG-uh-sawrs*) were mostly medium-sized dinosaurs with rows of bony plates or spines down their back. This built-in body armour made it very hard for predators to attack them. Stegosaurs walked on four legs and were herbivores, which means they ate plants instead of meat. **Can you count the spines on their backs?**

Hesperosaurus means "western lizard".

Hylaeosaurus means "woodland lizard".

Kentrosaurus grew to around 16 feet (5m) in length – that's about the same as 5 red kangaroos!

1 **Ankylosaurus**
(say: *ank-uh-lo-SAWR-us*)

2 **Dacentrurus**
(say: *da-sen-TROO-rus*)

3 **Emausaurus**
(say: *em-ow-SAWR-us*)

4 **Gastonia**
(say: *gas-TO-nee-uh*)

5 **Hesperosaurus**
(say: *hes-per-uh-SAWR-us*)

6 **Hylaeosaurus**
(say: *hy-lee-uh-SAWR-us*)

7 **Kentrosaurus**
(say: *ken-tro-SAWR-us*)

More stegosaurs

1 Tuojiangosaurus weighed around 3,300 pounds (1500kg) – that's three times as heavy as a grizzly bear.

2 Tarchia was a desert-dwelling dinosaur.

Here are some more stegosaurs. Bones from these dinosaurs have been found all over the world, but the most recent – and oldest set of bones discovered – was in Morocco in 2019.

Saichania means "beautiful". Can you see the large club on the end of its tail?

Minmi is named after a place in Australia called Minmi Crossing.

Scutellosaurus existed almost 200 million years ago.

Nodosaurus lived around 100 million years ago in North America and Canada.

1 **Tuojiangosaurus**
(say: *too-yang-uh-SAWR-us*)

2 **Tarchia**
(say: *tar-CHEE-uh*)

3 **Saichania**
(say: *sy-CHAN-ee-uh*)

4 **Minmi**
(say: *MIN-mee*)

5 **Scutellosaurus**
(say: *skoo-tel-o-SAWR-us*)

6 **Nodosaurus**
(say: *no-do-SAWR-us*)

Frills, horns and beaks

1 Anchiceratops was around 20 feet (6m) long – that's about 6 cricket bats.

2 Stenopelix existed in Germany around 127 million years ago.

3 Graciliceratops was tiny – it was only about 3 feet (80cm) long. That's about the size of an umbrella!

These dinosaurs are also herbivores, but they have bony frills around their heads and horns. They are called ceratopsians (say: *ser-uh-TOP-see-uns*), and many of them had jaws that were shaped more like beaks!

4 Styracosaurus means "spiked lizard".

5 Yinlong walked on two legs. Its name means "hidden dragon".

6 Nedoceratops had a beak that helped it to strip leaves from plants.

1 **Anchiceratops**
(say: *ank-i-SER-uh-tops*)

2 **Stenopelix**
(say: *sten-uh-PEL-iks*)

3 **Graciliceratops**
(say: *gras-i-li-SER-uh-tops*)

4 **Styracosaurus**
(say: *sty-ruh-ko-SAWR-us*)

5 **Yinlong**
(say: *YIN-long*)

6 **Nedoceratops**
(say: *ned-uh-SER-uh-tops*)

Small but speedy

1. Agilisaurus lived 150 to 170 million years ago in China.

2. Lycorhinus lived in South Africa and had long teeth.

These dinosaurs were called ornithischians (say: *or-ni-THISH-ee-uns*) – they were two-legged, fast-running plant-eaters with pelvic bones shaped like those found in birds. Ready, steady, go!

3 Heterodontosaurus's name means "different-toothed lizard".

4 Othnielia had a stiff tail that helped it to balance when it ran.

5 Lesothosaurus had five "fingers" on each hand.

1 **Agilisaurus**
(say: *uh-ji-li-SAWR-us*)

2 **Lycorhinus**
(say: *ly-co-RY-nus*)

3 **Heterodontosaurus**
(say: *het-ur-uh-don-to-SAWR-us*)

4 **Othnielia**
(say: *oth-NEE-lee-uh*)

5 **Lesothosaurus**
(say: *luh-soo-too-SAWR-us*)

Walking plant-eaters

1. **Parksosaurus lived in Canada around 75 million years ago.**
2. **Atlascopcosaurus came from Australia.**
3. **Hypsilophodon had broad, flat teeth.**

This group of herbivores that walked on two legs are called ornithopods (say: *or-NITH-uh-pods*). Some of them were pretty small compared to other members of this family (which you will see when you turn the page). Most of the dinosaurs shown here were around 3 to 13 feet (1 to 4m) in length.

4 Valdosaurus bones and teeth have been found around the globe in West Africa, the United Kingdom and Eastern Europe.

5 Yandusaurus had four toes on each foot and five fingers on each hand.

6 Zephyrosaurus means "west wind lizard".

1 Parksosaurus
(say: *parks-uh-SAWR-us*)

2 Atlascopcosaurus
(say: *at-luss-cop-kuh-SAWR-us*)

3 Hypsilophodon
(say: *hip-sih-LOF-uh-don*)

4 Valdosaurus
(say: *val-duh-SAWR-us*)

5 Yandusaurus
(say: *yan-doo-SAWR-us*)

6 Zephyrosaurus
(say: *zef-i-ruh-SAWR-us*)

Large ornithopods

1 People believe Iguanadons had very long tongues.

2 Parasaurolophus was about 36 feet (11m) long and weighed around 7,500 pounds (3500kg) – that's about the same weight as a hippopotamus.

These medium-sized herbivores usually walked or reached for high-up leaves on two legs. They would stand on all four legs when they were grazing for food on the ground. Almost 100 different types of ornithopods have been discovered so far. They became extinct around 66 million years ago.

Did they eat-it-all-osaurs

1 Dracorex has a dragon-like skull, and its name means "dragon king".

2 Prenocephale lived over 66 million years ago in Mongolia.

This group of herbivores had thick, ridged skulls and walked on their back legs. They are called pachycephalosaurs (say: *pack-i-SEF-uh-luh-sawrs*), which means "dome-headed". Researchers think some of them may have also eaten small animals and insects. (Creatures that eat both plants and animals are called omnivores [say: *OM-nuh-vawrs*].)

Goyocephale was around 7 feet (2m) in length and its name means "decorated head".

The first complete Stegoceras skull was discovered in 1924.

Homalocephale means "level head".

1 **Dracorex**
(say: *DRAY-co-rex*)

2 **Prenocephale**
(say: *pree-nuh-SEF-uh-lee*)

3 **Goyocephale**
(say: *goy-uh-SEF-uh-lee*)

4 **Stegoceras**
(say: *steg-uh-SER-us*)

5 **Homalocephale**
(say: *ho-muh-lo-SEF-uh-lee*)

Prosauropods

1 Twenty Riojasaurus skeletons have been found.

2 Mussaurus means "mouse lizard".

Sauropods were plant-eating dinosaurs with long necks and tails and small heads. These creatures, prosauropods (say: *pro-SAWR-uh-pods*), were their smaller ancestors who walked on two legs and had shorter necks. When you turn the page, you'll see how they grew!

3 Pantydraco bones were found in Wales.

4 Ammosaurus existed over 180 million years ago. Its name means "sand lizard".

5 Jingshanosaurus was one of the last prosauropods.

1 Riojasaurus
(say: *ree-o-huh-SAWR-us*)

2 Mussaurus
(say: *moos-SAWR-us*)

3 Pantydraco
(say: *pan-tee-DRAY-co*)

4 Ammosaurus
(say: *am-muh-SAWR-us*)

5 Jingshanosaurus
(say: *jing-shahn-uh-SAWR-us*)

Super-sauropods

1 Supersaurus was the longest dinosaur that ever lived, measuring 115 feet (35m) from head to tail. This is just a bit longer than 3 city buses end to end.

2 Pelosaurus lived in England.

3 Austrosaurus lived over 100 million years ago.

You can work out the names of some sauropods from the places where they roamed millions of years ago. **Can you tell which of these take their names from Nigeria, Malawi and Australia?**

Beware the titanosaur

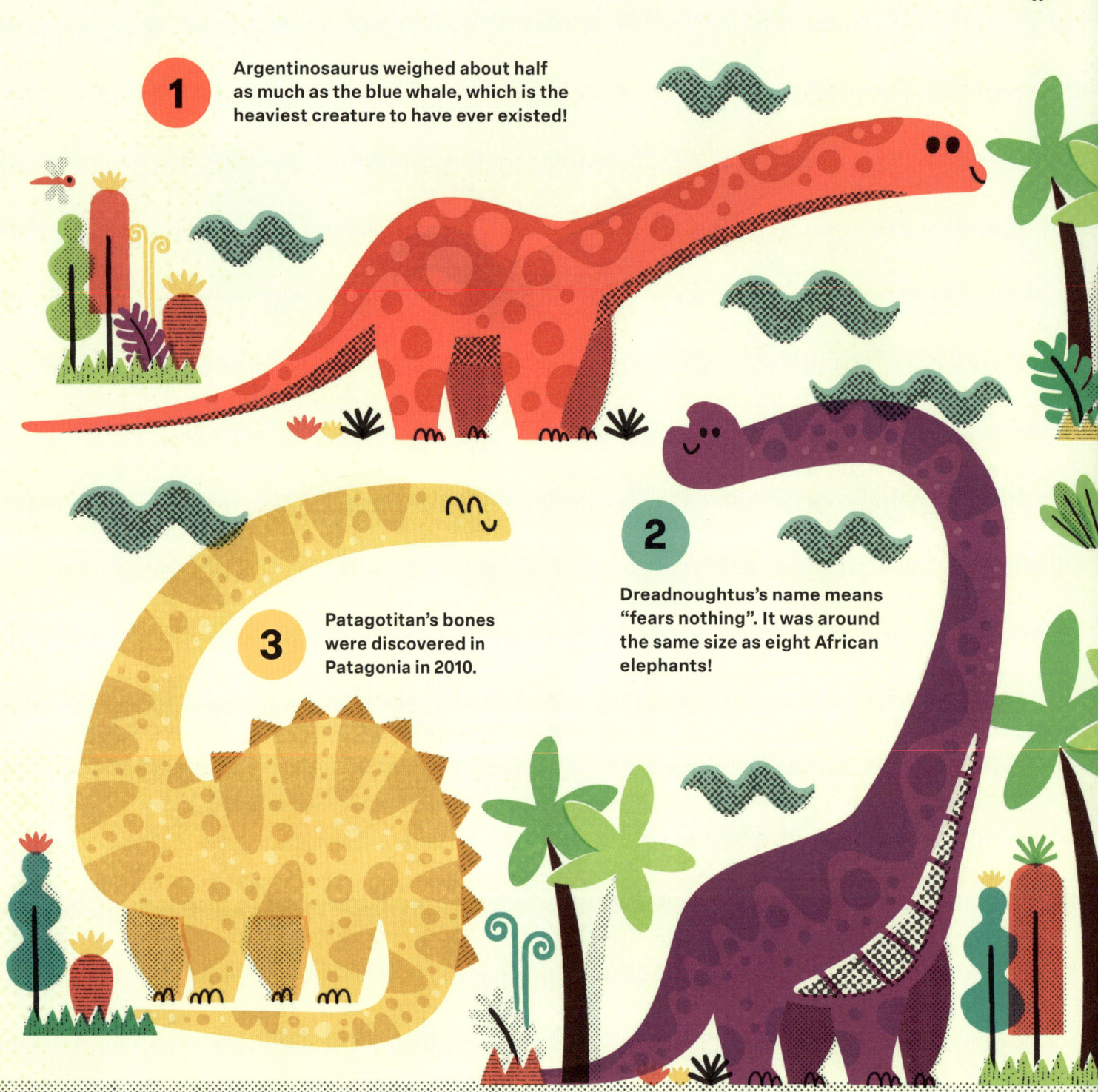

1 Argentinosaurus weighed about half as much as the blue whale, which is the heaviest creature to have ever existed!

2 Dreadnoughtus's name means "fears nothing". It was around the same size as eight African elephants!

3 Patagotitan's bones were discovered in Patagonia in 2010.

Titanosaurs (say: *ty-TAN-uh-sawrs*) are the largest dinosaurs in the sauropod group – and the biggest creatures to have ever walked on the surface of our planet. Don't get under their feet! The only place titanosaur fossils (the remains of prehistoric creatures found in the ground) haven't been discovered is Antarctica.

4 Pelosaurus means "monstrous lizard".

5 Saltasaurus was one of the last sauropods to have existed before they became extinct.

1 **Argentinosaurus**
(say: *ar-jen-teen-uh-SAWR-us*)

2 **Dreadnoughtus**
(say: *dred-NAWT-us*)

3 **Patagotitan**
(say: *pat-uh-go-TY-tun*)

4 **Pelosaurus**
(say: *pel-uh-SAWR-us*)

5 **Saltasaurus**
(say: *salt-uh-SAWR-us*)

Jaws, claws and teeth

1 Velociraptors may have been covered in fine feathers rather than scales and were the size of a turkey.

2 Deinonychus means "terrible claw".

3 Only a few Borogovia bones have ever been discovered, so nobody really knows what it looked like!

Carnivores, herbivores, omnivores – all of these dinosaurs walked and ran on two legs. They are called theropods (say: *THER-uh-pods*).

4 Troodons are thought to be one of the smartest dinosaurs.

5 Eoraptor lived 228 million years ago in Argentina.

6 Harpymimus lived 99 million years ago in Mongolia.

7 Sarcosaurus bones were discovered in England. Its name means "flesh lizard".

1 **Velociraptor**
(say: *vuh-LAHS-uh-rap-tur*)

2 **Deinonychus**
(say: *dye-NON-i-kus*)

3 **Borogovia**
(say: *bor-uh-GO-vee-uh*)

4 **Troodon**
(say: *TRO-uh-don*)

5 **Eoraptor**
(say: *EE-uh-rap-tur*)

6 **Harpymimus**
(say: *har-pee-MY-mus*)

7 **Sarcosaurus**
(say: *sahr-kuh-SAWR-us*)

Here comes trouble

1. Spinosaurus is thought to be the longest meat-eater of them all.

2. Megalosaurus was one of the first dinosaurs to be discovered.

3. Allosaurus had curved dagger-like teeth that stopped its prey from escaping.

The largest meat-eating dinosaurs were also theropods. **You've already met the Tyrannosaurus, but how many of these other ferocious beasts do you know?**

Becklespinax lived over 132 million years ago in the United Kingdom.

4

Saurophaganax grew to around 40 feet (12m) in length.

5

Only one, incomplete Deltadromeus skeleton has been found, in Morocco.

6

1 Spinosaurus
(say: *spy-nuh-SAWR-us*)

2 Megalosaurus
(say: *meg-uh-lo-SAWR-us*)

3 Allosaurus
(say: *al-uh-SAWR-us*)

4 Becklespinax
(say: *beck-uhl-SPY-nax*)

5 Saurophaganax
(say: *sawr-uh-fuh-GAH-nax*)

6 Deltadromeus
(say: *del-tuh-DRO-me-us*)

Early birdies

1 Archaeopteryx lived 147 million years ago, in the late Jurassic period.

2 Conchoraptor feathers have been found along with their bones.

Although not all of them could fly, some small theropods had feathers or wings. It took around 70 million years, but today's birds are descendants of this group. The word *raptor* also refers to a bird of prey such as an eagle or a hawk, which eat meat and have sharp beaks and claws. **Can you spot the feathers on these dinosaurs?**

3 Caudipteryx's tail ended in a fan of feathers.

4 Bambiraptor was named after the Disney film character Bambi.

5 Khaan's name means "ruler".

6 Microraptors only weighed around 3 pounds (1.3kg) – but they had four wings instead of two!

1 Archaeopteryx (say: *ahr-kee-up-TEH-rix*)

2 Conchoraptor (say: *KAHN-kuh-rap-tur*)

3 Caudipteryx (say: *kaw-dip-TEH-rix*)

4 Bambiraptor (say: *BAM-bee-rap-tur*)

5 Khaan (say: *KAHN*)

6 Microraptor (say: *MY-krow-rap-tur*)

Flying high

1. Pteranodons lived in large flocks and had furry bodies. Their name means "wings and no teeth".

2. Sinocalliopteryx lived 125 million years ago in China.

3. Ichthyornis was a seabird with a long beak.

The group of flying reptiles that existed during the time of the dinosaurs were called pterosaurs (say: *TER-uh-sawrs*). Not many fossils remain because their bones were fragile. However, we know some were as big as a fighter jet and others were the size of a sparrow!

4 Quetzalcoatlus is the largest flying creature ever known. It was as tall as a giraffe.

5 Iberomesornis may have been tiny, but it had a beak with teeth and claws on its wings.

6 Copepteryx were flightless birds that became extinct around 23 million years ago.

1 **Pteranodon**
(say: *tuh-RAN-uh-don*)

2 **Sinocalliopteryx**
(say: *sy-nuh-cal-ee-up-TER-ix*)

3 **Ichthyornis**
(say: *ik-thee-OR-nis*)

4 **Quetzalcoatlus**
(say: *kwet-zal-ko-OT-lus*)

5 **Iberomesornis**
(say: *i-beh-ro-muh-SAWR-nis*)

6 **Copepteryx**
(say: *cope-TER-ix*)

On the wing

1 Eopteranodon was small and toothless.

2 Phosphatodraco had a long neck and a wingspan of around 16 feet (5m).

3 Confuciusornis was the first prehistoric bird discovered with a beak and no teeth.

Pterosaurs had hollow wing bones that were light and flexible enough to allow them to lift off and fly. Their wings were made of skin that acted like a sail.

4 Pterodactyls lived at the same time as the Tyrannosaurus rex.

5 Aerotitan fossils were discovered in Patagonia in 2012.

6 Dimorphodon wasn't good at flying.

1. **Eopteranodon** (say: *ee-uh-teh-RAN-uh-don*)
2. **Phosphatodraco** (say: *fahs-fuh-to-DRAY-co*)
3. **Confuciusornis** (say: *kun-fyoo-shuh-SAWR-nis*)
4. **Pterodactyl** (say: *ter-uh-DAK-til*)
5. **Aerotitan** (say: *er-uh-TY-tun*)
6. **Dimorphodon** (say: *dy-MAWR-fuh-don*)

Below the surface

1 Eurhinosaurus had a sharp snout like a swordfish.

2 Ichthyosaurus had huge eyes to help it see in deep, dark water.

We've seen that dinosaurs once walked the Earth, but these large marine reptiles swam in its seas. These types of creatures were called ichthyosaurs (say: *IK-thee-uh-sawrs*), which means "fish lizard", even though some of them were around the size of a whale!

3 Temnodontosaurus had teeth strong enough to eat armoured prey.

4 Shastasaurus was 69 feet (21m) long but had no teeth.

5 Mixosaurus had a fin on its back like a dolphin.

1 **Eurhinosaurus**
(say: *yoo-ry-no-SAWR-us*)

2 **Ichthyosaurus**
(say: *ik-thee-uh-SAWR-us*)

3 **Temnodontosaurus**
(say: *tem-no-don-tuh-SAWR-us*)

4 **Shastasaurus**
(say: *shas-tuh-SAWR-us*)

5 **Mixosaurus**
(say: *mix-uh-SAWR-us*)

Plesiosaurs and mosasaurs

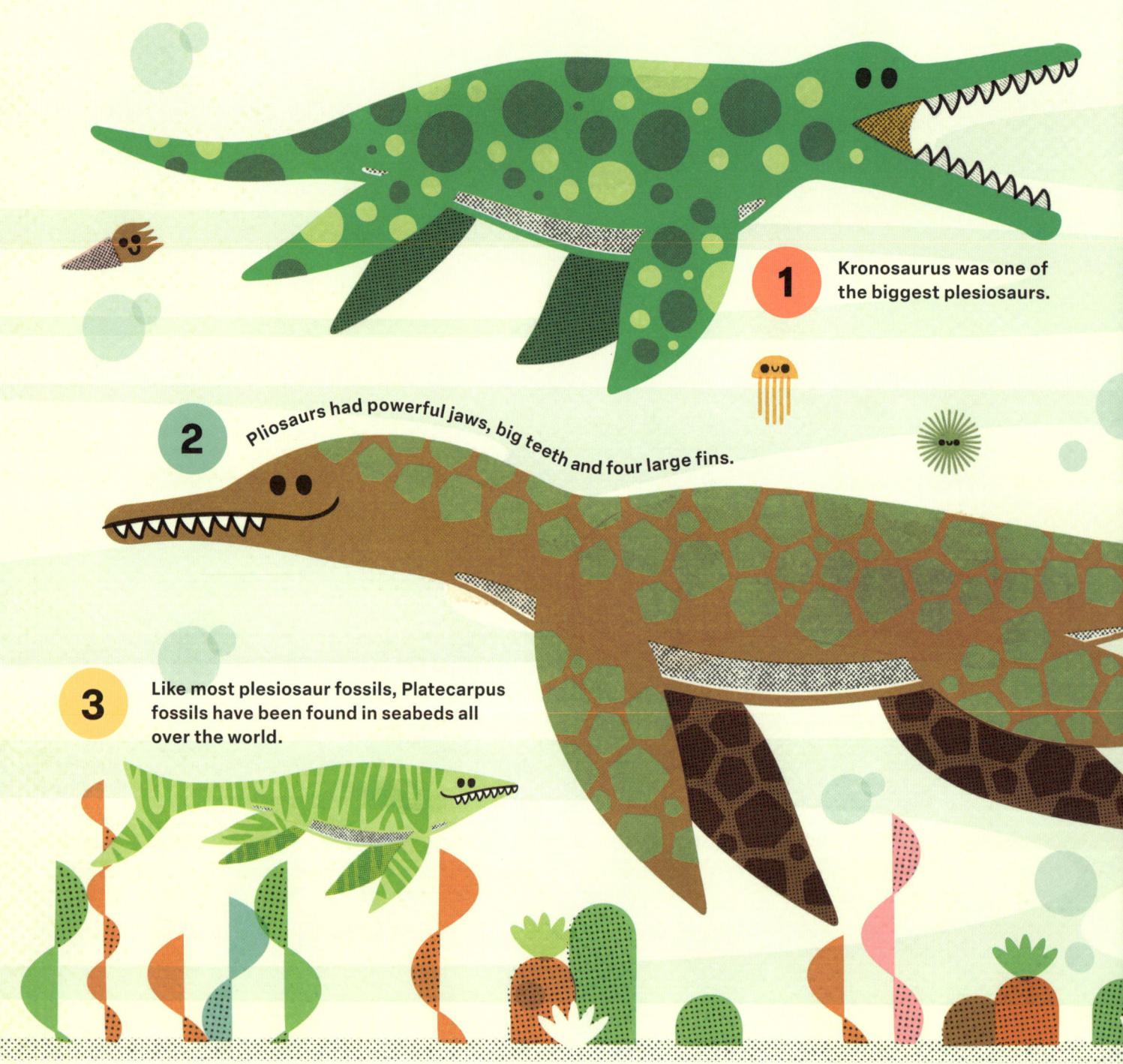

1. Kronosaurus was one of the biggest plesiosaurs.

2. Pliosaurs had powerful jaws, big teeth and four large fins.

3. Like most plesiosaur fossils, Platecarpus fossils have been found in seabeds all over the world.

Plesiosaurs (say: *PLEE-see-uh-sawrs*) existed alongside ichthyosaurs for around 100 million years. They had short tails, long necks and flat bodies, and some were 49 feet (15m) long – that's the length of 1 ½ buses!

Mosasaurs (say: *MO-suh-sawrs*) appeared around 5 million years after ichthyosaurs died out. For 30 million years, they were the oceans' top predators because they could grow to the size of a bus, had two sets of teeth and would eat pretty much anything!

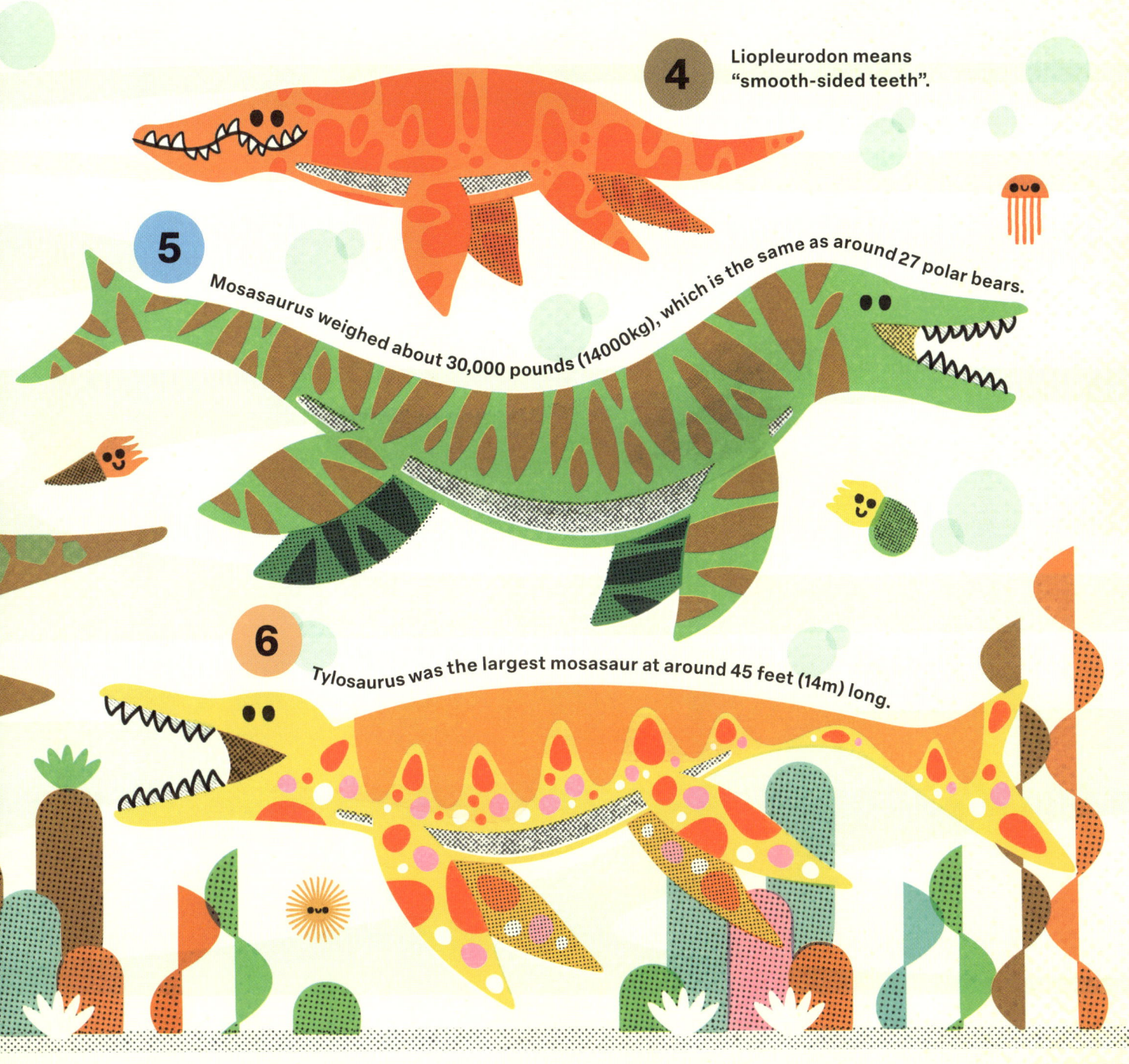

4 Liopleurodon means "smooth-sided teeth".

5 Mosasaurus weighed about 30,000 pounds (14000kg), which is the same as around 27 polar bears.

6 Tylosaurus was the largest mosasaur at around 45 feet (14m) long.

1 Kronosaurus
(say: *kro-nuh-SAWR-us*)

2 Pliosaur
(say: *PLEE-uh-sawr*)

3 Platecarpus
(say: *plat-uh-CARP-us*)

4 Liopleurodon
(say: *lee-uh-PLUR-uh-don*)

5 Mosasaurus
(say: *mo-suh-SAWR-us*)

6 Tylosaurus
(say: *ty-luh-SAWR-us*)

Mega-sharks and turtles

1

Great white sharks are believed to have lived around the same time as the megalodon. They are still around today, and even though they are much smaller than the megalodon, they are still big compared to a person.

Megaladon means "big tooth".

2

The megalodon was the largest shark ever to have existed and had teeth the size of a human hand. The female megalodon was larger than the male and could weigh around 50 tons (45t) – about the same as 22 great white sharks. Thankfully, this huge sea creature is now extinct.

Did you know that the last of the surviving prehistoric marine reptiles is the turtle? Despite swimming in our seas for millions of years, sea turtles are now an endangered species. It's up to us to protect this ancient animal from extinction.

3 The Leatherback turtle is the only species of turtle with tough skin instead of a shell.

4 The Loggerhead sea turtle has a large head and very strong jaws.

1 Great white shark

2 Megalodon
(say: *MEG-uh-luh-don*)

3 Leatherback turtle

4 Loggerhead sea turtle

45